CONTENTS

Introduction

Over the last few years, there has been a tremendous advance in the area of printing in textiles. From the conventional methods of printing, the industry is on for a revolution with a new sophisticated method of printing namely digital printing. In the textile printing area this technology has made great impact in the hard copy output from computer aided design (CAD) systems or in the quick reproduction of an existing design by color-photocopying.

Examining the production time scale of a textile print from design conception to a bulk print, considerable advantages have been gained from design selection and sampling using digital printing technology. To obtain a feel for the production time scale of a textile print let us examine the situation for a totally manual approach of a textile print. The figure shown below gives an overview of the scenario uptil a few years back for the majority of textile printers[1].

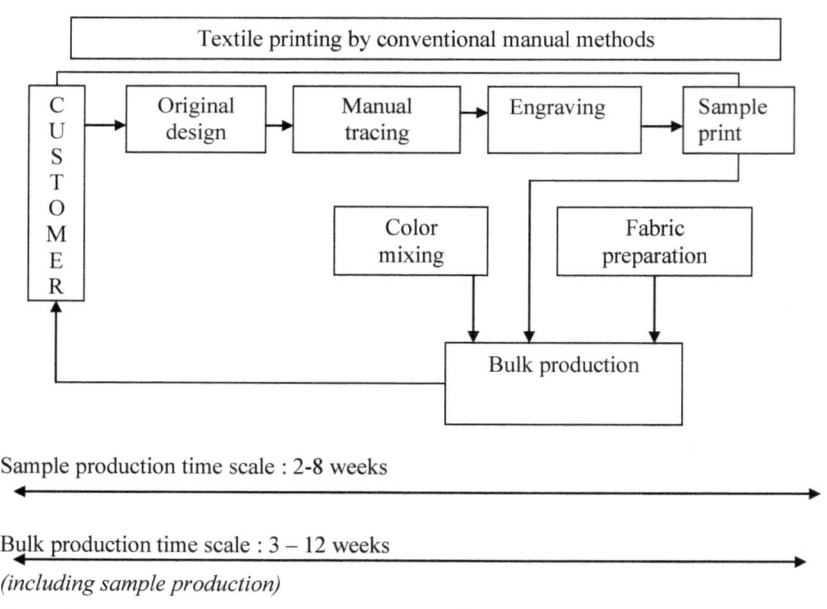

Sample production time scale : 2-8 weeks

Bulk production time scale : 3 – 12 weeks

(including sample production)

figure 1: textile printing by conventional methods[1]

The original design is manually traced, films produced and individual screens or rollers engraved for each color (i.e. if a textile design has 24 colors then 24 screens have to be produced unlike the CMYK system used in the digital printing). A sample print is then produced in a number of colorways on the textile substrate supplied by the customer for approval. After approval (following inevitable colorway changes!) the bulk printing takes place. The time scale indicated in figure are still typical of those currently achievable by the textile printer[1].

With the right investment the printer can scan designs into a CAD system where one can reduce the number of colors, manipulate designs, put into repeat, color and produce separations. The digital information produced can then subsequently used to produce screens directly through latest laser engraving technology, or by using computer produced films. The majority of the CAD applications start with at least an overall initial colored design which ,after being scanned into the system, can then be modified and colored. All this technology does is to speed up the overall printing production time scale.

With the rotary screen printing, the most popular printing technology at the present situation, set-up time is disproportionate to run time. The efficiency, the percentage of runtime to downtime, is 40 percent on a 4000m run and 60 percent on a 14000m run. But new developments like quick change of screens between runs, changing from one set of screens to another on the fly, plug and play controls and synchronisation are improving production efficiency. In addition printing speeds of around 85 meters per minute, are increasing.

On the other hand, set up times are minimal in digital printing although printing speeds are lower in the range of 20-25 meters per hour, while the pre- and post- treatments of fabric coating, mounting, fixing, steaming and washing all add to the production time. The technology is also eco-friendly with all the ink going on to the fabric and none to the waste[2]. The following are the available technologies for digital printing[2]:

1) Thermal drop on demand (DoD) ink jet
2) Piezoelectric DoD Ink jet
3) Airbrush/valve jet
4) Continuous ink jet
5) Electrophotography: Laser and LED

Image storage

An area of major concern to the digital photographer is the image storage. It concerns both the temporary storage of the image and then the subsequent final storage into the computer hard disk in the computer workstation and image manipulation. Internal storage of images is done by using EPROM microprocessors (erasable programmable memory) which is a ROM chip to which data is written once. This data is the read and/or erased. The image data remains permanent even when the power supply is switched off.

At present there are three key technologies vying for the market share in the removable, peripheral storage media sector.the first of these is the PCMCIA cards (personal computer memory card international association), available as type II or type III configurations.the maximum storage capacity is currently 50 Mb for type III configuration and 260Mb for the type II.

Other memory cards include IBM microdive, Lomega's click and the Sony company's memory stick flash. The Microdive system, a mini hard drive which plugs directly into a compact flash type II socket, uses GMR (giant magnetoresistive) head technology which allows impressive image data transfer rates of 30.1 to 45.2Mb per second.

Lomega Click is another disk-based system which uses Winchester technology for storage, whereas the Sony memory stick is another flash memory card which can hold up 64Mb of image data[3,4].

Computer-related removable storage media

Before the advent of multimedia based printing and digital photography, the floppy disk was the simple way of processing word data. However, because the image files are far greater in size, new removable storage systems have been incorporated with either magnetic or optical data storage technologies.

Most recent magnetic storage device include Lomega Zip and Jaz drives. These systems are relatively inexpensive. The systems are easily portable and can be used heavily for business purposes.

Equal is the popularity of optical storage compact disks. This has come about due to the availability of inexpensive compact disk (CD) writers. The compact disk recordable (CD-R) and the compact disk recordable-writtable (CD-RW) and Kodak photo CD are most important for digital photography[3,4].

The latest optical disk to enter the market is digital versatile disk (DVD), which has been developed specifically for the distribution and playback of images. As dyes tend to be used as the recoding media (IR absorbing organic dyes, mainly cyanine) for all optical data storage systems, therefore much of research activity is on specific dyes for the DVD systems. One of the major breakthroughs which has the capacity to pack the image data more closely than is possible by optic or magnetic media. In theory this could lead to storage volumes up to a million times greater than those obtainable by CDs.

Charged coupled device (CCD)

The image capture onto the computer hard disk can also be done by using charge coupled device (CCD).This is done by using charge-coupled device (made of a semi-conductor).These devices have light sensitive surfaces, containing hundreds and thousands of picture elements, known as pixels the light sensitive silicon semiconductor converts the light energy falling on it, to an electrical potential. for a typical digital conversion the following sequence of events take place.

a) The semi-conductor material responds to light energy, with each individual pixel accumulating electrons in proportion to the light received.

b) Each of the individual pixels is then allotted a value which depends on both its voltage level and its position within the CCD array. Digital photography uses a binary number system consisting of zeros and ones. These digits are often referred to as bits, the greater the number of bits available, the greater the number of values. A digital system requires a brightness resolution of 24 bits, thereby allowing any combination of 256 shades of red, green, and blue to be ascribed to a single pixel, having a possibility of over 16 million different colors [3] *(fig 2)*.

Scanners

Scanners are peripheral devices that convert images into digital data The development of the scanner has proved to be one of the most significant advances in reprographics technology. The first two scanners were invented by Murrray and Morse in 1941 and Hardy and Wurzburg in 1948.The resolution of the scanner is directly proportional to the number of photosensitive picture elements present. Thus, for a typical flatbed scanner with a scan width of 8.5 inches and 2450 pixels, the maximum resolution available will be 300 dpi approximately. However most of the scanners incorporate an interpolation function which doubles the resolution [5,6].

Scanner types:

There are basically four different types of scanners available. These being the hand-held, the flatbed, the film and the drum scanner. The scanner types are dealt below [5,6].

a) **Hand-held scanners**

There are two types of hand-held scanners, the first being the hand driven type which operates by moving the scanning head across the image. The second type is the window type which functions by means of placing the window frame over the image, with the linear array scanning head being drawn automatically across the image area within the window frame. The output resolution for this type of scanner varies between 100 to 400 dpi approximately. The important disadvantage of this type of scanner is that it is incapable of being able to scan images much larger than enprint photograph size.

b) Flatbed scanners

CCD based flatbed scanners are the most popular image capturing method. Flatbed scanners incorporate a linear CCD array scanning head which contains several CCDs arranged in a line on a long silicon chip. The original reflective image is illuminated from below by a fluorescent or a halogen lamp. Certain models involve individual colors (red, blue, green) sources while others use a single white light with equivalent filters. There are two ways of relaying the colored information from the image to the computer, these being the single pass mode and the three pass mode. For the single pass mode the color image is scanned only once by the scanning head as the lighting alternates between the red, green and blue lamp source. The color information is read simultaneously as a line. In the three pass mode the color is scanned three times i.e.once each with the red light, the green light and the blue light respectively.

As the full color image is read simultaneously as a line, the specific color and intensity of light falling on each individual CCD element creates an electrical charge on the element proportional to that level.this electric charge is an analogue charge which is then systematically passed along chains of cells to an analogue digital converter (A/D), where it is converted into a digital data.the now clear CCD is ready to receive the next light induced charge *(fig 4).*

Film scanners

The design of a film scanner is specific to film format images, which can be in the negative or positive (transparency) form. The majority of the systems use 35 mm size file format, usually with an APS (advanced photo system) option. The professional models, such as the Kodak professional RFS 2035 plus, are capable of scanning colors or monochrome images (negative or positive) at a range of resolutions. These are as follows with their respective file sizes along in the brackets:

1) 1333 dpi (8Mb)
2) 1600 dpi (11.5 Mb)

3) 1800 dpi (14.6 Mb)

There are a number of advantages associated with scanning film as opposed to reflective prints, these being as follows:

1) Speed and cost advantage due to not having to print the image in a photographic paper.

2) No optical data loss from a print production process.

3) Both color and monochrome images can be produced from color negative or positive images.

4) Film images can be scanned either mounted or unmounted.

Drum scanners

These scanners are capable of converting both reflection and transparency images into electronic data, prior to high quality reproduction drum scanner operates by means of a photomultiplier tube (PMT). The PMT, together with a light source, operate either side of hollow perplex tube into which the original image is attached. The PMT is mechanically moved over the image whilst the drum spins at high speed (1000 to 1600 rpm).A high quality drum scanner can scan up to 12000 dpi without interpolation. The *fig* illustrates the principle of a typical drum scanner. The xenon light source is focussed by means of a lens system on to an area of the original image. Light from a small section of the image then enters the sensor unit and is directed on to a series of dichroic mirrors. Reflected light from the mirrors then pass through the red, the green, and the blue filters on to the photo multiplier tube. Following amplification, the analogue signals are converted into digital data by means of an A/D converter*(fig 5)*.

Image file formats

Images that has been digitized, either by original capture using a digital camera or converted capture (from a silver halide original) by a scanner, must then be stored within a file type, the file format, which defines how the image is stored in the file and subsequently how the information will be displayed on the computer monitor or hardcopy output device. There are currently fifty different graphic file formats in existence and, therefore, awareness of the principal ones available is a must for every printer[3,6].

Usually image file formats comprise of two parts, the file header and the image data. The header would be expected to contain the following information:

a) Horizontal dimensions of the image (in pixels)

b) Vertical dimensions of the image (in pixels)

c) Image data type (greyscale, color)

d) Image bit depth

e) Compressing technique (if used)

    The header information contains all the requisite information to construct the original data. In many cases the image data, prior to storage, will have been compressed. Again the file header is responsible for its reconstruction. Compression of image files are required to reduce to a more manageable size, enabling either storage or transmission of the image.

Common image file formats

    The most commonly available image file formats are described as follows[3,6]:

a) TIFF (tagged image file format).

    This is the most common format for Image saving, and was developed jointly in 1986 by Microsoft and Aldus. TIFF is an uncompressed file format whereas TIFF with LZW (Lempel-Ziv-Welch) is a lossless compression method, whereby the image data is compacted and the file size reduced, without any loss of image detail. This format is favourable for digital imaging printers.

b) JPEG (Joint Photographic Experts Group).

    This is the name devised by the committee that developed this image file format. The format is based on discrete transform algorithm (DCT). This undertakes the task of analysing either 8 by 8 or 16 by 16 pixel arrays within an image, and performs an averaging procedure for each cell. This results in the significant decrease in the image size. JPEG compression is lossy and is capable of compressing images to one hundredth of the original file size. The image that has been jettisoned cannot be retrieved back. It is not advisable to compress a JPEG file for a second time as it reduces image quality further. A new file format namely JPEG2000 is wavelet based compression method. It provides a number of advantages over the DCT method of compression.

c) EPS (Encapsulated Postscript)

    This format is device independent and, therefore, images can be readily transferred between applications. It is capable of storing either bit-mapped images or object oriented graphics. EPS uses the postscript language code to define images with vectors, which link the image resolution to the print resolution. Two problems associated with the EPS format are the large file size and the fact that the files can be printed only on a postscript compatible laser printer.

d) *IVUE*

The IVUE format is used within the high end image manipulation program Live picture for display and image processing, with the object of speeding up these processes. Files are converted usually, from either photoCD or TIFF format to the pyramid type structure of the IVUE format, by Live picture software. Once manipulation is complete the file must be converted into its original format for output purposes. An advantage of the IVUE format is that they can be imported into the complementary Adobe Photoshop.

e) *PICT*

This is a general format file for Apple Mac computers and is used for storing, either object oriented or bit-mapped images. The current format supports 8 bit color depth (256 colors). PICT files can be readily converted to JPEG files by using Apple Mac Quicktime, thus enabling insertion of images into many common window applications eg Word and Photoshop.

e) *PCD (Photo Compact Disc)*

This is a proprietary file format, developed by Kodak, for storing Photo CD images which has been encoded in the YCC color model and then compressed. Each individual PCD contains five files of the one image at different resolutions. This format is mainly used for digital printing.

f) *Photoshop*

This is a native file format of the industry standard image manipulation program Adobe Photoshop. The format retains information relating to the likes of the masking channels, which may well be lost when saved in the other file formats. The format is famous in digital imaging outlets. Photoshops are used extensively for digital printing and high end digital photography.

Image compression

Image compression is required to reduce most image files to a more manageable size, enabling either storage or transmission of the image. This aspect of digital imaging is an area of current research. There are currently three basic types of compression methods available i.e. lossy, lossless and visually lossless. The lossless type of compression allows an exact reconstruction of each of the individual pixel values and is usually referred to as image coding. Huffmann and LZW codes are the two widely used compression algorithms for the lossless coding of images[6].

If some loss of the pixel values can be accepted, very much higher compression ratios for the images can be realised. This type of compression is known as lossy and is attained by the use of transform methods that relate in principle to the JPEG standard. Both the chrominance (color information) and the luminance (greyscale image) are compacted separately. The reason is that the human eye is more tolerant in the assessment of color as opposed to the grey scale information. Thus the loss of color information is more acceptable than the loss of greyscale, which is required by the eye to indicate boundaries.

The final type of image compression is visually lossless, which is the method utilised by the Kodak photo CD system. The format has one luminance (Y) and two chrominance (C1/C2) components. The first of the two chrominance components is unique to the red-green balance, whereas the second is unique to the yellow-blue balance. The system relies on the limitation of human vision, with respect to color, enabling the removal of a percentage of this particular information whilst retaining the more important greyscale information.

Digital imaging software

Following the digitalisation of the photographic image, it is possible to manipulate the images in various ways, using the personal computer containing a specific manipulation software program. There is now a proliferation of software packages available. Some of them are PC specific and some of them Apple Mac specific[3].

Image manipulation

There are a number of image manipulation software packages available for image manipulation. Many of the less sophisticated, or abridged versions of a more complex original, are included with the likes of the color printers and scanners. Some examples of manipulation packages for PC and/or Apple Mac workstations, are as follows.

a) Adobe Photoshop

b) Sartori PhotoXL

c) Live Picture

d) XRes

e) Picture Publisher

f) Photodeluxe

Manipulation software includes image adjustment, color adjustment, image repairing and retouching. Image adjustment includes the ability to crop, size or re-orient the image. Tonal adjustments, too, can be made, involving the setting of the shadow and highlighting points. Also classed as a basic effect is the Unsharp mask filter. Although this is a digital effect, the term, itself, relates to the reprographic film process whereby a similar result is obtained by sandwiching an original positive image next to the light, soft, unsharp negative image, during exposure. For color adjustment of the image, a key component of the program is the hue/saturation command, which is essentially a version of the CIELAB system, which, itself, is derived from the Munsell system. The CIELAB system uses three spatial coordinates a^* (red-green axis), b^* (yellow-blue axis), L^* (lightness axis) for brightness control[6,7].

When the hue/saturation command is chosen one can alter hue, saturation, and brightness components of the image, both completely and selectively. The ultimate, though, in precision control, with respect to the image, comes with the selective color command. This enables the color balance to be finely adjusted in both the additive and subtractive primaries and in the whites, blacks and neutrals.

For repairing/retouching of an image, usually the principal route is by the rubber stamp tool, which is used to clone parts of an image. The rubber stamp is a brush tool and like equivalent brush tools the size, opacity, and shape can be altered for specific requirements.

.

Advanced manipulation techniques

The Adobe Photoshop software package contains many techniques which could be classified as advanced. These include filters, layer effects and black and white effects. One of the reasons why Adobe Photoshop has stayed ahead of its main competitors, is its wide spread support of third party filters eg Kai's power tools. The filter is already built into the program, prior to any new ones being added, are numerous and include impressionistic or mosaic effects, addition or reduction of noise levels, application of lighting effects, and image distortion[3].

Ink jet printing in textiles

After the CAD manipulation stage it is customary to obtain a pre-print assessment via a hard copy printer. Up to a few years ago CAD systems incorporated one of the many non-impact printers available. The actual choice depended on the price level but was normally based on ink jet printing technology (bubble jet printer) or a thermal transfer printer. A textile hard copy would be more useful in design interpretation and this has led in the past to many ambitious prospects aimed at developing such output based on jet printing technology[1].

Definition of types in ink jet printing
Coarse resolution

Maximum resolution 40 lines/inch and based on valve control technology. The valve control technology can be direct or indirect (deflector). There are commercially two main commercially available systems. The Milliotron system uses an array of jets with continuous streams of dye liquid which can be deflected by a controlled air jet. The chromjet from the Austrian company Zimmer uses computer base on/off valve systems to control the flow of the dye liquid. There is also an electromechanical valve system which is computer controlled to open and close rapidly so that the liquid is fired in succession of short pulses.

The above mentioned printers are called carpet jet printers and the resolution is relatively coarse with the maximum resolution of 40 jets per inch which is unacceptable in the textile printing field.

The majority recent projects developing ink jet technology for textiles have looked essentially at adapting computer ink jet printing technology rather than valve technology developed for carpets[1,8].

Fine resolution

Resolution of up to 300 lines/inch with a certain ink jet printers are possible. Fine resolution printers can be split into two basic technologies.

a) Drop on Demand (DoD)

b) Continuous stream (CS)

Within these two technologies there are a number of sub-groups. It is in this fine resolution area that there has been the most recent research in textile printing area. The whole area of fine resolution ink jet printing has recently been extensively reviewed. In the textile printing area there are essentially two ink jet technologies of interest[1,8].

Continuous Stream ink Jet printing

In the continuous Ink Jet system, ink is forced at a high pressure through a small jet (nozzle). The emerging stream of ink is broken into small droplets. By selectively charging these droplets they can be deflected when passing through high deflection plates. There are two possible methods of obtaining a design by this process. In the first process the uncharged droplets from image and the droplets are deflected to waste. This is known as 'Binary jet stream'.

In the second method the charged droplets are deflected onto the substrate in a pre-determined manner and the uncharged droplets are collected in the gutter. This is referred as 'raster scan method.'

Both systems are generally based on Ink Jet printing technologies developed by Sweet in 1964.The binary method being further developed by Professor Hertz and his group at Lund University[1,9].

Drop-on-Demand Ink-Jet Printers

This technology produces an ink droplet when required and fires this onto the substrate. The ink droplets are not charged. The DoD printers fall into two broad based classes[1,9].

a) Those systems that produce a drop using a piezo-electric transducer.

b) Those that use thermal excitation (the bubble jet type) to produce a drop.

Piezo-electric DoD Technology

Piezo-electricity is a phenomenon of producing electricity by application of pressure on a crystal which is capable of conducting electricity through it. This method has been suitably exploited for ink jet printing technology.

A piezo-electric crystal contracts when a current is applied. Piezo ink jet printing relies on different principles for the expulsion of ink from the cartridge nozzles. With this technology, an electrical charge is applied to the cartridge nozzles and excites a small piezo crystal that is inside. When the piezoelectric crystals are stimulated, the crystals change shape and squeeze the ink chamber. This action is similar to the action of squeezing an oil can, and forcefully expels the ink from the nozzle tip.

Since the piezoelectric process does not utilize heat, the cartridge life of these printers is greatly expanded, cartridges should last a minimum of one year under heavy usage. Piezoelectric printheads can use a wider range of inks than thermal inkjet printers because the heat is removed from the process. This means that solvent-based ink systems and pigmented-ink formulations will be more readily available, which increases the development capabilities for better inks in the future. Although piezo is currently the lesser-utilized technology, many experts predict that the long-term development of ink jet print devices will use the piezo technology because of the greater through-put speeds offered and the wider latitude with the types of inks that can be developed[1,9].

Thermal ink jet system

Thermal ink jet printing, also known as "Drop on Demand", employs a process of super-heating the ink inside the print cartridge to about 400 degrees. In this method a tiny heating element is incorporated behind each individual nozzle. When a current is applied the temperature of the resistor rises very quickly to 400 degree celcius. As the ink heats up, vapo bubbles are formed inside the cartridge, which expand, explode, and then force ultra-fine droplets of ink out of the printhead's micron-size nozzles and onto the media. As the ink leaves the nozzle head, it creates a vacuum that pulls in fresh ink. This process is repeated thousands of times per second.

A key feature of thermal inkjet printers is that like desktop models, they use disposable print cartridges, usually one for each of the four process colors: (Cyan, Magenta, Yellow, black). The cartridge contains both the ink supply and the printhead. Typically, these cartridge-driven units produce excellent print quality at 300 to 600 dpi, but with slow print speeds of approximately 10 to 20 sqft /hour. Thermal inkjet printers are popular choices for corporate art departments, design studios, quick-printers, sign shops, photographic labs, and screen printers that hat expect to produce only a handful of prints per day[9,10].

Laser systems

Currently laser technology is used in a variety of ways to produce hardcopy output images from digitally stored data. The most commonly used laser technology is elecrophotography, which covers the two common output technologies of laser printing. The major stages for image development and transfer, with respect to laser printing are as follows:

a) The photoconductor drum receives a uniform electrostatic charge[3].

b) The laser writes the image information on the photoconductor drum, at the same time as the electrostatic charge dissipated.

c) The result is a latent electrostatic image both in the uncharged image areas and the charged non-image areas of the photographic drum.

d) Then the latent image is made visible by the application of a toner, which is repelled to the uncharged image areas, because the particles have the same charge as the non-image areas.

e) Finally the toner-filled image areas are transferred to the oppositely charged fabric surface.

It is important that color laser printers contain four drums and store image colors in cyan, magenta, yellow and black elements. There is a wide variety of consumable materials associated with laser technology. These include photoconductors, charge-generation materials, charge transport materials, electron transfer material, the developer and the charge control agents. Although photoconductor materials were originally inorganic eg zinc oxide, doped selenium and amorphous silicon, now due to their wider spectral response, either pigments, or dyes, of which typical examples are azo pigments, phthalocyanines, perylenes and squaryliums. Charge transfer, which are divided into, either 'hole transport'or 'electron transport' categories, include both triarylamines and hydrazones. The components involved in the developer depend upon whether the developer is a liquid or a solid system. A liquid system comprises a toner dispersed in an insulating solvent, whereas a dry system comprises of a toner/carrier combination. The toner itself is made up of two components, a thermoplastic resin and a coloring agent. The resin accounts for between 40 and 95% by weight of toner particles, whereas, the coloring agent comprises between 5 and 10 %. The principal function of the resin is to physically fuse the image to the paper by the application of heat or pressure. The commonly used resins for toners are styrene acrylics and styrene butadienes. With respect to the coloring agents, pigments tend to be used more commonly than the dyes, with color image obtained by subtractive color mixing, using cyan, magenta, yellow and black coloring agents. Some of the pigments used most commonly to give the full color image are as follows:

a) cyan-phthalocyanines (C.I. pigment blue 15:1)

b) magenta- basic dye complex pigment (C.I. pigment violet 1)

c) yellow- diarylide yellow (C.I. pigment yellow 12)

d) black – carbon black (C.I. pigment black 6/7)

The initial charge-control agents were designed specifically for negative charging toners, these being metal azo-complexes (for black toner system only). However, for color copying the current trend is toward colorless materials. Thus, colorless negative charge control agents tend to be metal-complexes or salts, usually formed from non-colored precursors eg salicylates. Colorless positive charge control agents tend to be quaternary ammoniun salts or related pyridinium salts.

Inks used in digital printing:

Fabric, unlike paper, is a three dimensional structure and the ink and colorant requirements vary over a large range. Practical limitations exist on the range of fabrics and colors that can be produced with a single ink set. On some fibers that are absorbent, like wool and cotton, the ink is absorbed quickly and easily, so bleeding of the water-like ink jet ink is minimized even without a pretreatment. Unlike the thick, paste-like ink used in conventional screen-printing, these water-like inks will bleed badly on non-porous fibers like polyester and nylon. A mechanism to control bleeding must be incorporated to avoid the ink wicking along the non-porous fibers of the textile. This also is important in applications that require print through on the design to give nearly equal color on both sides of the fabric. In traditional printing this is controlled by the high viscosity of the inks used. With ink jet printing pre-heating the textile or addition of a fabric pretreatment may help control these effects. The binding mechanism of the pigment to the textile and the reaction of the dyes with the fibers usually require a complimentary pretreatment chemistry and/or post treatment to achieve the optimum result. The bottom line is that the ink, textile and the printing system must be designed to control bleeding while achieving the hand, correct color and fastness required by the intended application[2,11,12].

The operative here is "intended application." Printed textiles are sold to many different market segments for a variety of end uses, including fashion textiles, home textiles and soft signage (flag and banner). The target market and end use will ultimately determine the fabric, ink and post processing requirements.

Digital inks include liquid and dry toner electrographic and magnetographic ink-jet inks, including water based dyes, solvent based dyes, pigmented solid phase wax, UV cured and dye sublimation and dye diffusion systems. Other components present are mineral oils, hydrocarbon resins. Toners are usually produced by each machine manufacturer and vary in composition but generally consist of pigment, base binder resin, modifier resin and charge control agents agents include iron oxides, Cr(III) or Co(III) complexes of azo dyes, salicylates, nigrosines and cetylpyridiniumchloride depending on whether toners are negatively or positively working. Other additives include slip agents such as polypropylene, wax or silicones. The major set back of these digital ink systems are predominantly fused toner systems which are difficult to de-ink.the indigo electro ink has a polymerising vehicle. There is a significant difference in the applicability of digital inks to the normal screen printing inks. The characteristics include properties like ink viscosities, thickeners used, colorants, purity of the inks and the pre-treatments and the post-treatments. There are significant differences in the application of various chemicals[12,13,14].

Ink characteristics for conventional screen printing and digital printing

Ink characteristic	Screen printing	Ink-jet printing
Viscosity	High	Low
Thickener	Used	Not used
Colorant	Excess	Minimal
Pre & Post-treatments	Required	Required
Purity	Impure	Ultra pure
Pigment type	Many	Restricted

Digital inks can further be classified into solvent based inks and aqueous based inks. In solvent based inks the dissolving media is an organic solvent whereas in an aqueous based ink the dissolving media is water.

Solvent based inks

Ink chemistry	Fibres	Post-processing	Markets supported
Dye	Polyesters	None	Soft signage
pigment	Vinyl,Polyester, Nylon	None	Outdoor signage

Fabrics which are made out of polyester and other synthetics like vinyl and nylon are normally not given any post-treatments. The dyes are pigment have no affinity for the polyester fibre and are simply impregnated into the solid matrix of the fabric. Pigments are so chosen that it does not cause blooming effect. Blooming results in loss in color value in a fabric like polyester.Great care has to be taken to print a polyester fabric.The system is highly complex and any deviation from the normal parameters can be fatal[11,13].

Water based inks

Ink chemistry	Fibres	Post processing	Market support
Acid	Silk,nylon,wool	Steam/wash,can be Drycleaned	Fashion textiles,indoor soft Signage
Disperse Dyes (Sublimation)	Polyester	Heat fixation	Fashion textiles, Indoor & outdoor Soft signage
Reactive dyes	Natural fibres, Cotton, silk, Rayon,wool.	Steam /wash	Fashion textiles, Indoor soft Signage.
Direct dyes	All fibres	Steam/wash	Fashion textiles
Pigments Without Binder	All fibres	Dry heat	Indoor and Outdoor Signage
Pigments With binder	Cotton and Possibly polyester	Dry heat	Indoor and Outdoor soft signage

The colorants used for ink-jet printing can be either made up of

a) Pigments

b) Dyes

Pigments

Pigments are differentiated from the dyes as coloring matters in the basis of their solubility characteristics. Pigments are insoluble in the medium in which they are incorporated by dispersion and they remain as solid discrete particles held mechanically within the fabric matrix. Pigments thus tend to resist dissolution in solvents which they come into contact in application to minimize problems such as bleeding and migration. In addition to solvent resistance, pigments are required to be fast to light, weathering, heat and chemical to the degree dependent on the demand of the particular application. For most of the application, it is the optical properties of the pigment which are more of importance. The most obvious of these is the ability to impart desired color. Pigments can further be classified into organic and inorganic pigments. Pigments are crystalline materials manufactured in a finely divided form. Solid state structure of pigment exhibits polymorphism i.e. capable of existing in forms with identical chemical composition but different lattice structures. Treatment of the surfaces of the particles is commonly used to improve the performance of the pigment. For example, use of organic surface active agents may lead to improved dispersability in the application medium. While inorganic material such as silica improves light fastness and chemical stability. The following are the basic requirements for digital printing with pigments[2,12,15].

a) Reliable, non-clogging
b) Depth of shade and brightness
c) Good fastness to crocking, washing abrasion
d) Soft hand
e) Cost
f) Total applicable system cost (printer, software, pigment inks)

Desirable pigment ink characteristics for digital printing

a) Purity: A high degree of purity will help minimize clogging of the print heads, the most common problem associated with ink-jet printing.
b) Particle size: As pigments are coarse and insoluble in nature, they must be finely ground in order to obtain the proper sub-micron size as well as particle size distribution.
c) Viscosity: Low viscosity, almost water thin, is necessary to allow easy flow through the micron size nozzles used in piezo Drop on Demand and thermal printing heads. This property seems less critical on continuous ink-jet nozzles.

d) Conductivity: Electrical pulse determines drop delivery and therefore the selection of the pigment ink components contributing to ink conductivity level is critical.

e) Surface tension: Drop formation and stability can be affected by surfactants and other components used in ink jet formulations.

f) Shelf life: Dispersion stability can determine reliability of the pigment being applied. Unlike most dyestuffs, pigments are not water soluble and can settle over time, causing printing malfunctions. Dependable storage stability over a period of time has been another difficulty associated with pigment inks[12,16].

Inks which contain a pigment, particularly with a binder also in the system, it is desirable to use printers of the piezo type. The reason being that the high temperature which is obtained by the tiny bubble creating components in the printhead has a marked effect on the ink. The high temperature causes charring which is technically termed as kogation[17].

In the case of pigment printing, special pigment binders such as those based on polyurethane dispersions are used rather than the conventional polyacrylates or acrylamides or butadienes. The advantage being that it has a less tendency to clog the jet orifices.

Dyes

Textile dyes are used in the form of their aqueous solutions. Dyes are required to have a high degree of solubility in water or in certain cases like vat dyes where it is capable of making it into water soluble form. Dyes are used as opposed to pigments because of the brighter colors that are obtained along with the wider color range. While in the application of the dyes on the fabric, the dye must be readily absorbed on to the fibre surface of the textile material, whilst the solvent must evaporate or penetrate the body of the fabric. The majority of the black, yellow and magenta dyes used in the ink formulation for printing are based on the azo chromophore.Usually black dyes are large aromatic compounds containing diazo, triazo, tetrazo chromophoric groups[11,12].

The most commonly used yellow chromophores are azopyridones, azopyrazolones, which exist extensively in hydrazo tautomeric form, due to the presence of an internal hydrogen bonding. This however makes them more feasible for photo decomposition. The most commonly used magenta chromophores are azo dyes based and is obtained through H-acid and γ-acid. H-acid dyes are bright and have moderate light fastness. Moderate light fastness is because of the hydazo tautomeric form.γ-acid dyes exist in azo form and as a result has a better light fastness[18].

Cyan dyes are based on copper phthalocyanines. These class of dyes show excellent photostability on a wide variety of textile substrates. There can be certain chemical modifications done in the phthalocyanin structure to shift the absorption band in the UV-visible range.

The physical form of the dye on the surface of the substrate is of key importance. The ideal solution is for the dye molecules to be aggregated into nanocrystallites, which ensures that the molecule at the surface will protect those trapped within the crystal[19].

Dye requirements

The dyes requirements for digital printing of textiles is vigorous. The dyes must satisfy the following criteria in order to be used in for printing using ink-jets:[13,20,21]

a) It should give a narrow absorption curve and it should strictly have no secondary absorption peaks.

b) It should have a high level of purity, strictly no electrolyte or insoluble matter.

c) It should possess high optical density.

d) It must have good thermal stability (for thermal and bubble jet printers).

e) It should be non-toxic (it should be Ames negative i.e. non-carcinogenic).

As printing technology advances, progress must and will be supported by ink chemistry improvements. The important trend to follow is the development of pigment systems or alternative chemistry for textiles. Pigment systems have not been easily adapted for the textile inkjet environment and early introductions have been criticized for color brilliance and fastness. For the sewn products industry, the pigment trend is significant for a couple of reasons. In contrast to dye based systems in which dye class must be matched with fiber type, pigments are substrate independent. They can be used for printing a wide range of fibers and fabrics including blends. This will have a great deal of interest for producers of home furnishings, bed linens, and certain apparel products. In addition, pigments and alternative chemistry that do not require steam fixation will simplify the path between printing and cutting. A dry fixation unit could potentially be mounted between a printer and single-ply cutting unit. In contrast, the steaming process required by dye-based systems presents a barrier to integration. Steaming requires a separate process and may cause changes in dimension and shape that make part recognition more difficult during the cutting procedure.

Pre treatments and Post treatments

The treatments given to a fabric both before and after a ink jet printing process plays a pivotal role in the determining the efficiency and the quality of print obtained. The treatments are however complicated and are often employed in ink jet printing technology. Printing with a reactive dye and an acid dye based inks involves pre and post treatments in order for the dye to fix onto the fabric. This consists of a number of steps with a good degree of complexity. However printing with a pigment or a disperse dye based inks are usually simple and easily executed[17].

Reactive dye based inks are often used in ink jet printing technology mainly due to the ease of getting them into suitable ink formulations due to their water soluble ability. Reactive dye interacts with the cellulose to give rise to a covalent chemical bond. These bondings give rise to the high degree of wash fastness. In order to maximize the reaction between the reactive unit of the reactive dye and the cellulose molecule, alkali and heat are required. In ink jet technology, the alkali is applied by a pretreatment process as or else it interferes with the reactive dyes and the nozzle components if present the ink itself. The heat is applied after the printing process is over by using either a steam source or hot air fixation process. A separate wash process must be used to remove the unfixed reactive dye present superficially over the fabric[22].

Acid dyes are normally used to print silk, wool, or other polyamide fibres. A pre treatment is generally necessary to prevent wicking of the ink in the fabric. A separate post-treatment is given like steaming to fix the color and a separate wash off process is done to remove the unfixed dye.

Disperse dyes are the main printing colorants for polyesters. High temperature steaming generally used online to fix the dye after direct printing in an easy single stage process. The colored pigment is bound to the substrate with the help of a binder system. The binder is present either in the ink or is applied after the printing is over. The application route does not employ a pre-treatment or a washing process and is shorter than the reactive based inks.[23-25]

Formation of patterns in ink jet printing

The genesis of the individual droplets at the jet outlets follows a remarkably similar sequence and leads to the formation of circular drops on the substrate. As the jetted stream leaves the jet orifice it forms a tail which, if the conditions are ideal, eventually collapses into the head of the drop which becomes spherical before hitting the substrate surface. In reality there is a tendency, particularly with the bubble jet systems, for the long tail of the drop to form small satellite droplets but this does not necessarily affect the printed design.[17,26]

The volume of the individual drops varies considerably (from 50 to 1000 picolitre), depending on the type of machine and the duration and voltage of the driving pulse. As the depth of the shade increases, the colored spots increasingly intermingle on the surface of the substrate. Each printed pixel of a design, when examined under a microscope, can be seen to be composed of individual drops of the colored inks. The overall shade observed is dependent on the proportion of each of the primary colors applied and the nature of the substrate and the pre-treatments given.

Within the printer lies both the hardware and software to interpret the pattern data as it is downloaded. Each individual pixel element of the original design is printed as the number of colored spots arranged in a superpixel usually forming either a 4 by 4 or 8 by 8 matrix. Most jet printers used for textiles print at a definition of either 300/360 or 600/720 dpi. High definition printers, giving 1200 dpi or greater and often capable of printing with very small volume drops, are used more in the reprographics industry. The pixels per inch determine the design definition. Thus two printers, one of which uses a 4 by 4 matrix at 360 dpi and the other an 8 by 8 matrix at 720 dpi, will exactly give the same contour definition (90 ppi) despite the greater number of the individual drops of the ink projected onto the substrate for any particular depth of shade. This is illustrated in the figure for prints of neutral grey on Epson printer.

Attainable color gamuts

The range of shades which can be displayed by a VDU is very extensive; wider than can be achieved by any class of dye and very wider than is possible using only yellow, cyan, and magenta jet inks. The color characteristics of the different makes of the VDU do vary to some extent, mainly in the choice of the green phospor and this is particularly the case with the proposed SMPTE-240M standard for high definition televsion. Even standard monitors can display very much brighter shades in the red and the blue regions, yellows can be of similar brightness whilst in the cyan/bluish green region CMYK prints are marginally brighter [17,27](fig 5).

Limitations in the attainable shade range, which was formerly common with the earlier jet printers using only CMYK ink, have been overcome in some machines which now have a facility for using additional spot colors[28] (bright reddish-yellow, orange, red, deep blue and green).

Variables affecting shade reproducibility

In the textile ink jet printers are used either for sample approval prior to printing the pattern conventionally or directly for small scale production prints. The reproducibility of the print will depend on the following possible variations in[17] :

a) The substrate and how it is prepared.

b) The inks and the consistency with which the ink jet system delivers them to the fabric surface.

c) The absorption/fixation characteristics of the dye used in the inks.

d) The reliability of all the jets in the printhead to remain functioning.

Effect of substrate variation

Depending on the type of pretreatment given to the fabric, the color yield can vary between 50-100 %. Almost all textile fabrics are given a pre-treatment with a thickener containing *inter alia* humectants, alkali or acid and surface active agents, depending on the dye/fibre system. Such a pretreatment (i.e. kezlan plus fused silica) can have major effect on the color yield of the print resulting in spots that are more sharply defined, which particularly affects heavy shades[17] *(fig 6)*.

For optimum color yield the spots of ink within each superpixel must remain clearly defined on the fibre surface because once the white fibre surface is completely covered by the ink spots there is no appreciable increase in the depth of the color print. Thus it is not desirable for the ink-droplets,, once they land on the fibre surface, to spread but at the same time the inks must be absorbed sufficiently on the fibre surface and then allowed to dry so as not to cause smearing or marking-off faults. The table shows the effect of cotton fabric pretreatment on the surface wetting properties. The longer wetting time shown by the fabric treated with thickener and fuming silica is the result of the greater surface absorption of the liquid at the point of impact of the drop with a consequent reduction in wicking, i.e. exactly the conditions required for maximum color build-up in ink-jet printing. Each and every fabric type has its own chemical interaction with the surrounding chemical atmosphere and behaves in a different way from the other textile substrate. This gives rise to variation in the printability of the fabric. These can to some extent be overcome by giving some pre-treatments to the fabrics before they are subjected to printing[11,17],

Effect of cotton fabric pre-treatment on surface wetting properties

Preteatment	Wetting time (s)		Comparative color print strength
	Warp	Weft	
None	1.00	1.30	100
Kezlan	1.30	1.50	125
Kezlan + fumed silica	4.60	9.00	160

Once the spots of inks have dried in the fabric surface the dyes are fixed either by steaming or thermofixation.

Problems in printing

The most common faced in ink-jet printing is the impregnation of a relatively large volume of the fabric from only minute bubbles of ink. Moreover the ink has to be thin enough to avoid clogging of the nozzles but not so free flowing that it spreads too far across the fabric and loses definition of the basic design. Pretreatment provides a receptive surface for ink-jet printing and prevents unwanted penetration or spread of the ink. The technology also has to work on a wide range of materials, each with its own ink compatibility characteristics. Fibres can be synthetic or natural. The surfaces can be stretchable, flexible, often highly porous and textured. The inks should have good fastness properties, withstand finishing operations outdoor use with heavy wear, abrasion and cleaning. While the print has to look good, the ink should not alter the hand of the fibre. All the above are not satisfied by fabrics printed with digital printing. There is always some sacrifices made on one of the components in order to satisfy the other[11,12].

The major problems encountered while digital printing are as follows

1) As the single color drops, it must be placed on the fabric at exact spots. Maximum preciseness is required for the printing machine manufacturer.

2) There is a strong dependence on dye stuffs selection and the fabric structure

3) High resolution (upto 760 dpi) is required to avoid color areas from appearing granular or pixel like. The resolution value however does not represent the delicateness actually achieved on the fabric, as several color drops together form superpixels.

4) Higher resolution leads to lower output speed

5) Penetration varies depending on how many process colors are used

Conclusion

By limiting wet post processing, ink jet pigment printing has the potential to make "agile manufacturing" much more attractive. "Agile manufacturing" refers to an integrated, on demand, order and fulfillment process that includes the textile printing and product fabrication manufacturing processes. To implement "agile manufacturing" one must have the ability to print, cut, sew and ship immediately on demand. This capability can dramatically change the way sewn product and other printed textiles are produced.

Freedom from the requirement of using wet chemicals along with "agile manufacturing" will facilitate "distributed printing". "Distributed printing" refers to a small textile fabrication facility that receives the design and product information electronically, then produces product at or near the retail outlet. With these capabilities, along with the digital design, the **potential cost savings in the supply chain and the reduction in inventory and design risk**, the availability of digital ink jet pigment printing should drive conversion of some parts of the textile printing industry away from conventional screen printing.

Another key driver is printing speed. While all printers available on the market to date have been too slow to support mass production speed requirements, two recent product launches from BMT Technologies (dye sublimation/heat transfer printing) and Dupont boast digital textile printing speeds that can match the speeds of conventional rotary screen printing. While increasing speeds will certainly increase the likelihood of adaptation by the conventional printing industry, this solution merely supports the application of technology to existing business models, and provides no incentive to develop the mass customization business opportunities that the technology will support.

The greatest challenge to digital textile printing adaptation is the conventional textile printing industry itself, which continues to place analog restrictions on digital printing. They are looking at this new technology through the existing workflow and output of a 30 year old printing technology and trying to replace these analog processes and products with the digital ones. The result is a mind set that is focused more on what the digital technology can't do than what it can do

Implementing digital printing effectively means rethinking the overall system. It means designing different types of products that leverage the advantages of digital printing. It means sales and marketing strategies that leverage short run production. It means thinking about your business from the perspective of mass customization as opposed to mass production.

The single most costly element in today's soft goods business is the holding of product and parts in anticipation of a sale. This process supports the maxim that the best way to create profit is to mass produce and discount the surplus. To make this gamble work the apparel delivery system has developed a cumbersome structure designed to stockpile the inventory in staged production surpluses, then to sell finished product in tiered discount and therefore reduce the risk for each participant. While this process has in fact lowered cost, it has also lowered profit and customer choice. The solutions to profit erosion and lack of consumer choice are the same individualized mass production and delivery. Digital textile printing offers the technology to deliver mass customized product.

And therein lies the heart of the problem .. as well as the opportunity. Screen printing was developed in the industrial age where the economies of scale demanded a mass production business model. The technology itself, with the tremendous time and cost of screen development, demanded large production runs. The economic models of the conventional printing industry are based upon the volume of yardage that can be printed each year. Digital printing, in its current state of development, does not support this business model.

However, when looked at from the perspective of the wide format printing industry, an industry which is based on short run, customized production, textiles represent nothing more than a new media and a new market. This industry doesn't look at digital textile printing from the perspective of what it can't do, but rather all of the new markets it will allow them to enter. The challenge to these people is in learning the textile industry, whereas the challenge to the textile industry is in learning both the digital printing technology as well as a new business model under which to sell their products.

Bottom line, digital printing for short run production requires thinking outside the box. It requires an entirely new approach with new products and new marketing strategies that leverage not just printing technology, but CAD systems, cutting systems, information systems and even the Internet. Since the traditional industry has repeatedly demonstrated its lack of desire to change its ways and its fear of new technology, digital textile printing applications are as likely to come from outside the industry as they are from the inside.